The Cotton Gin

GREAT INVENTIONS

The Cotton Gin

MILTON MELTZER

BENCHMARK BOOKS

MARSHALL CAVENDISH
NEW YORK

Benchmark Books
Marshall Cavendish
99 White Plains Road
Tarrytown, NY 10591-9001
www.marshallcavendish.com
Copyright © 2004 by Milton Meltzer

All Internet sites were available and accurate when sent to press.

Library of Congress Cataloging-in-Publication Data
Meltzer, Milton, 1915-
The cotton gin / by Milton Meltzer.
p. cm.—(Great inventions)
Summary: Explains the mechanics of the cotton gin, invented in the late
eighteenth century by Eli Whitney, and describes how it enabled, tragically,
the vast expansion of the American slave trade.
Includes bibliographical references and index.
ISBN 0-7614-1537-8
1. Cotton gins and ginning. I. Title. II. Series: Great inventions
(Benchmark Books (Firm))

TS1585 .M45 2003
633.5'156—dc21
2002015308

Photo research by Candlepants Incorporated

Cover photo: Corbis / Bettmann

The photographs in this book are used by permission and through the courtesy of:
Corbis, 2, 24, 42, 45, 47, 48, 49, 52, 53, 56, 65, 74, 76, 87, 88, 96–97, 100, 106–107;
Bettmann, 16, 18, 34, 36, 38, 50, 66, 79, 80, 85, 93; David H. Wells, 17;
Underwood and Underwood, 68–69. Art Resource, NY/Erich Lessing, 10.
Art Archive/Geographical Society Paris/Dagli Orti, 12.
Hulton Archive by Getty, 14, 20, 26, 29, 31, 61, 62–63, 72, 83, 94, 102–103, 104–105.
National Archives, 40. Woolaroc Museum, 91.

Printed in China

1 3 5 6 4 2

C O N T E N T S

Foreword   9

One   Cotton, Cotton Everywhere   11

Two   From Farm to Factory   15

Three   Slavery—and Freedom?   25

Four   Mr. Whitney Can Make Anything   30

Five   Inventions: Good for Whom?   39

Six   Labor in the Mills   46

Seven   Picking Cotton   60

Eight   On the Auction Block   75

Nine   "I Have a System to Destroy"   82

Ten   The Trail of Tears   90

Eleven   Civil War—and After   95

Further Reading   111

Web Sites   113

Bibliography   115

Index   121

The Cotton Gin

Cotton—and the cotton gin. Familiar terms we rarely think twice about.

Cotton is all around us. Our body touches it daily. We wear it, sit on it, walk on it, sleep on it, shelter under it. The list of things made out of it goes on and on: jeans, dresses, shirts, blouses, socks, stockings, bathrobes, sheets, blankets, carpets, rope, nets, tents, bandages, towels, tea bags, uniforms.

The word *cotton* seems to have originated from the Arabic word *gurton*. What is it? To be technical for a moment, cotton is "one of a variety of plants of the genus *Gossypium*, belonging to the malvaceae family, native to most tropical countries." It's a shrubby plant, and in the Tropics can grow as high as twenty feet (six meters). When it is cultivated, however, it may be only a foot (thirty centimeters) high.

As for the cotton gin (gin being short for engine), it is linked forever to the name Eli Whitney. When that young man left New England and headed south in 1792, he had no notion that within months he would invent a machine that would change the course of history.

For that achievement Whitney is listed among the great American inventors—Alexander Graham Bell, Thomas Alva Edison . . .

This book looks into the world of Eli Whitney. It seeks to learn what he did and why he did it. And to explore his invention's profound and enduring significance, not only for America, but for the world beyond.

A WOMAN IS SPINNING COTTON IN THIS STONE RELIEF FROM SUSA, IRAN, CARVED AROUND THE EIGHTH CENTURY B.C.

Cotton, Cotton Everywhere

The story of cotton goes back thousands of years to the time when some peoples began domesticating plants and animals for food and clothing. Crops and livestock not only satisfied hunger, but yielded valuable materials, especially cotton—a fiber crop ideal for keeping people warm or shielding them from the sun.

Early farmers, both in the Old World and in the Americas, chose several species of cotton for long lint, which they used to weave textiles. Other species of cotton furnished fiber for woven clothing in China, Central America, India, Ethiopia, central and southern Africa, and South America. Each species was native to that particular part of the world.

Pieces of cotton cloth more than 7,000 years old have been found in prehistoric caves in Mexico. Five thousand years ago Egyptians in the Nile Valley were already wearing cotton clothing. Around the same time, people living along the Indus River in what is now Pakistan were growing cotton and weaving it by hand. In the ancient world of Greece and Rome, the historian Herodotus and the naturalist Pliny mentioned cotton in their writings. Around A.D. 700, a Chinese emperor, on the occasion of his ascending to the throne, wore a cotton robe. Poetry of his era celebrates the beauty of the cotton flower. Much later, in 1516,

THIS 1774 ENGRAVING SHOWS A MAN CARDING COTTON IN THE MIDDLE EAST.

there is documentary evidence that the people in southern Africa were growing cotton and wearing cotton clothing.

If you watch cotton being cultivated, you will see how it produces creamy-white flowers, which soon turn deep pink and fall off. This leaves the small, green seedpods, called cotton bolls, which contain the seed. Growing from the outer skin of the seed are seed hairs, or fibers. These are tightly packed within the boll, which bursts open upon maturing, revealing the soft masses of the fibers. These are white to yellowish white and about three-fourths of an inch (two cm) to twice that in length.

There are three large groups of cotton fibers (or staples), classified by length. The first group, rated highest in quality, includes the Sea Island, Egyptian, and pima cottons. These long-staple cottons are the scarcest and the hardest to grow. They are used for fine fabrics, yarns, and hosiery. The second group is the standard medium-staple length, often called American Upland. The third group is the short-staple, coarse cotton. It winds up in carpets and blankets, in cheaper fabrics, or as blends with other fibers.

Today 95 percent of the cotton in the world belongs to the medium-staple length, the cotton species *Gossypium hirsutum*.

SIR RICHARD ARKWRIGHT, INVENTOR OF THE WATER-POWERED SPINNING FRAME

From Farm to Factory

In early colonial days, cotton was not a major crop, because converting the white bolls into clean lint for the making of thread was a slow and tedious job. But cotton cloth was highly valued, so the colonies grew a little for their own use, and housewives mixed it with wool for spinning. Some cotton was shipped abroad to England because King George III accepted cotton as payment of rent for lands owned by the crown.

The long-staple cotton known as Sea Island entered the American South from the Bahamas toward the end of the American Revolution. It was produced only in a small area along America's southeastern coast. Because its smooth seed did not stick to the lint, cleaning the cotton was not a big problem.

Whether in England or America, spinning cotton thread and weaving it into cloth were vital parts of almost every family's daily life. In the early days of the Industrial Revolution, two inventions speeded up the process, each of them small enough to be used in the home. But the invention, by the Englishman Richard Arkwright, of a large spinning frame that required power moved the spinning process from the home to the factory.

Eng.d by Wm N. Dunnel, N.Y. from a Painting by Cole, Boston.

SAMUEL SLATER, FOUNDER OF NEW ENGLAND'S EARLIEST TEXTILE MILLS

At first, horsepower drove Arkwright's machine, and then, in 1769, water-wheels did. Ten years later, Samuel Crompton's "spinning mule" improved the quantity and quality of what Arkwright's machine was capable of. All this, before steam became the power source in the factories.

The mechanization of industry had a double effect in England. On the one hand, it forced workers out of the home and into what the poet William Blake called the "dark Satanic mills." But on the other hand, lowering the price of cottons to equal that of woolens provided washable clothing and diapers that improved the personal hygiene of mother and child, so that infant mortality was sharply reduced.

America's first cotton mill was built in Massachusetts in 1787. But in three years, it failed. Its machinery was not up to producing

A Slater cotton mill on the Blackstone River, in Pawtucket, Rhode Island

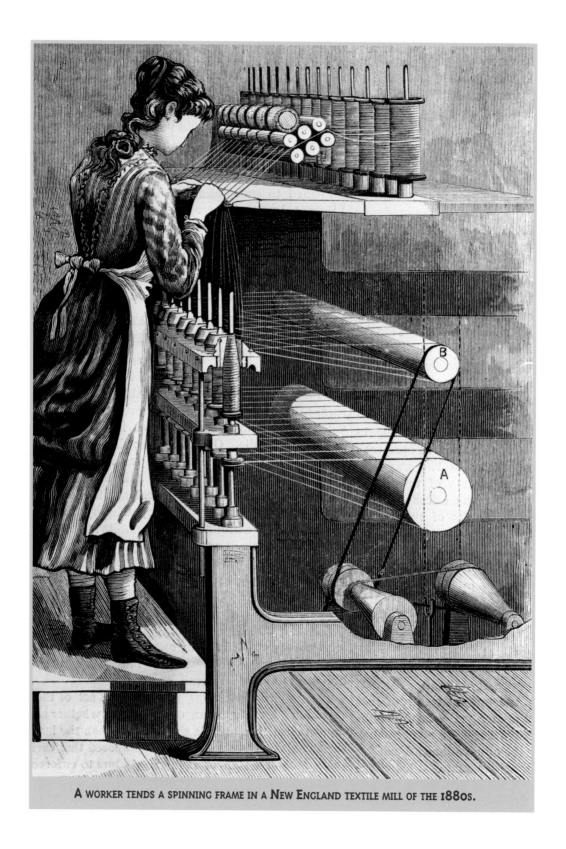

A WORKER TENDS A SPINNING FRAME IN A NEW ENGLAND TEXTILE MILL OF THE 1880S.

the quality of goods the British were turning out. But how could the Americans catch up with them? The British government allowed neither textile inventions nor textile mechanics to emigrate. One young man, Samuel Slater, ignored the law. In 1789, he emigrated from England to America with his head full of exact memories of how every machine in the factory he'd worked in operated and how it was repaired.

Told of a mill recently built in Rhode Island by Moses Brown, a Quaker merchant who was eager to find a manager who knew how to run it, Slater wrote to Brown—and got the job. When Slater inspected Brown's factory, he found it good for nothing. Although he had brought no drawings with him, Slater built from memory a series of machines based on the designs of the British inventors.

Slater's first mill in Pawtucket opened late in 1793. Business was slow for quite some time, but Slater kept at it, and eventually more mills were set up in New Hampshire and Massachusetts.

The Slater machines were simple to operate. Just as was done in England, Slater staffed his factories with children

For Great Britain the cotton industry with its extensive system of textile mills helped spread the nation's influence, as well as its rapidly growing empire in the early nineteenth century. The empire is long gone, but the legacy of the red-brick factories and the workers laboring at the machines remains.

Richard Arkwright, the industrial pioneer, first used rivers and streams to power his new textile machines in 1771 in the village of Cromford. Cotton spinning moved from being a cottage industry performed at home to a booming factory business. Communities were rapidly transformed.

In 1785, the banker and entrepreneur David Dale introduced a radical new concept in attempting to change the face of factory towns. He founded a model community on the banks of the river Clyde in central Scotland. His son-in-law, Robert Owen, developed it into what became the internationally famous town of New Lanark. The community featured free medical care, reduced working hours, a village discount store, and a school for the "formation of character" in what Owen modestly called "the most important experiment for the happiness of the human race at any time in any part of the world."

A few of these textile towns, considered models for their time, are now listed as World Heritage Sites by the United Nations Educational, Scientific, and Cultural Organization (UNESCO). They appear on the same list as such better-known World Heritage Sites as the Pyramids of Giza in Egypt and the Taj Mahal in India.

CLOTH MADE AND PRINTED BY THE

MERRIMACK MANUFACTURING C.º

LOWELL, MASS.

INCORPORATED 1822.

Warranted Fast Colors.

THIS LABEL TELLS THE BUYER THE CLOTH WAS MADE IN THE MERRIMACK FACTORY IN LOWELL, MASSACHUSETTS.

from four to ten years of age. He broke with English harshness by giving the children good food and humane treatment. Remember, this was a time when American children were set to work on the family farm as soon as they could walk.

Mitchell Wilson, the historian of American science and invention, assessed what Samuel Slater did:

> Slater neither invented anything, nor improved what he brought here; but he was the first in this country to set up a system of manufacture in which the successive steps of the skilled artisan were broken down into such simple components that a group of children could outproduce the finest craftsman. It was the one system ideally suited to a country that was to be plagued by a shortage of skilled manpower for another seventy-five years. No one saw any discrepancy between such a system and the American goal of enhancing the dignity and human value of the individual. The American factory fed, clothed, and equipped men for the fight against the hostile universe; and the factory system was actually considered to be a victory for the American creed of freedom.

Slater wasn't the only man to copy the designs of British textile machines. In 1813, an American, Francis Cabot Lowell, returned from a two-year stay in England with secretly made sketches of British textile machinery driven by waterpower. Securing a huge investment, his company built a textile factory in Waltham, Massachusetts, using waterpower and based on Lowell's revised and improved designs.

The factory was so great a success that in 1823 the company added another plant, built in the new town of Lowell, Massachusetts. (It was named in honor of Francis, who by then had died.) Within a short time, the plant was turning out millions of yards of cloth per year. Lowell had become the textile center of America.

By 1850, Lowell's population had swelled to 32,000 people. The workers tended 30,000 spindles and 9,000 looms. The production rate of the mills was higher than that of any other factory in the world.

In its early years, Lowell became famous for its special workers. They were mostly young girls recruited from farm families. The girls looked on work in the Lowell mills as the chance to broaden their experience and to learn about the world beyond their village. Authors, scientists, and educators made a point of lecturing in Lowell on their speaking tours. The girls lived in comfortable dormitories, where standards of good behavior were maintained.

In 1834, the British author Harriet Martineau visited a textile factory and wrote:

> Five hundred persons were employed at the time of my visit. The girls earn two, and sometimes three, dollars a week, besides their board. The little children earn one dollar a week. Most of the girls live in the houses provided by the corporation, which accommodate from six to eight each. When sisters come to the mill, it is a common practice for them to bring their mother to keep house for them and some of their companions, in a dwelling built by their own earnings. In this case, they save enough out of their board to clothe themselves, and have their two or three dollars a week to spare.
>
> Some have thus cleared off mortgages from their fathers' farms; others have educated the hope of the family at college; and many are rapidly accumulating an independence. . . . The people work about seventy hours per week, on the average. . . .

By today's view, seventy hours a week is hardly paradise. Yet in that early period of industrialization, the paternalism of the Lowell mill owners was widely admired. As other managers took over, their orders were to increase production and to cut costs in order to match the competition and improve profits. The number of looms each worker had to tend was increased from one to three or even four. (Speedup, it's called.) Owners paid bonuses to supervisors if they got more work out of the employees.

When America's first great depression struck in the 1830s, conditions rapidly worsened. The Lowell "utopia" was dying. Some visitors to Lowell then spoke of the mills as "factory prisons." In 1845, testifying before a state legislative committee, textile workers declared they were confined "from 13 to 14 hours per day in unhealthy apartments" and were thereby "hastening through pain, disease and privation, down to a premature grave." They ended their testimony by calling for a ten-hour day. The names of girls who protested bad work conditions were sent around to other companies. Such blacklisting spread widely to other managements who made sure no "agitators" found work.

While this was going on in the textile mills of the North, what was happening to the men, women, and children who were raising the cotton in the South?

THIS PRINT DEPICTS THE CULTIVATION OF COTTON ON A SOUTHERN PLANTATION.

Slavery—and Freedom?

The men, women, and children who were raising cotton in the South were black. And they were slaves. It was their hard labor that supplied the cotton for the textile mills in the North and in England.

You probably know something about how it happened that plantation labor in the American South was rooted in slavery. Slavery itself is almost as ancient as humankind. People of every color enslaved one another in prehistoric times and have continued to do so on down through the centuries. It was in the early 1600s when slave ships began to carry Africans to the New World. Some 12 million people were victims of the forced migration from their native lands to North America as well as to the Caribbean and to South and Central America.

As staple agriculture—such crops as rice, indigo (a plant used to make blue dye), tobacco, sugarcane, and cotton—began to spread in the colonies, more and more slaves were imported to do the work. When the American Revolution erupted in 1776, the slave population had climbed to half a million. Only a small number of slaves were brought to the northern colonies because they had no large-scale farming calling for

JOHN QUINCY ADAMS, THE SIXTH PRESIDENT OF THE UNITED STATES, WHO WAS A POWERFUL VOICE AGAINST SLAVERY AS A CONGRESSMAN IN THE 1830S AND 1840S

mass labor. The great majority of slaves was bought for labor on southern plantations.

How could slavery be reconciled with the Revolution's cry for freedom, liberty, and equality?

Out of the Revolution came the world's most enduring democratic republic. The Founding Fathers signed the Declaration of Independence, which declared that "all men are created equal." Yet the southern delegates to the Philadelphia convention that shaped the Constitution were almost all slaveholders. Their fortunes, their prestige, and their political power rested upon the institution of slavery.

The Constitution that the delegates worked out sanctioned and protected slavery, without ever using the word. Representation was apportioned on a three-fifths basis for "other persons" (slaves); the slave trade was extended for twenty years; and provision was made for the return of fugitive slaves.

This outcome was the product of bargaining between the economic and political interests of the northerners and the southerners. Accepting slavery troubled the conscience of some on both sides. But what eased their minds was the racist excuse so widely accepted for slavery—that white people were superior to people of any other color and therefore had the right to rule over them.

The states that make up what is now called the Deep South threatened not to accept the Constitution unless northerners supported the three-fifths clause. That clause counted only three-fifths of the slaves in each state for the purposes of congressional representation and in the electoral college.

No one spoke out more powerfully in the Congress against slavery than John Quincy Adams (1767–1848). Elected the sixth president in 1824 and defeated for re-election, he was asked by his Massachusetts constituents to represent them in Congress in 1831. (No other president has ever served thus.) He attacked all measures that would extend slavery, and in 1833, speaking on the floor of the House of Representatives, he pointed out that the three-fifths clause

> has secured to the slaveholding states the entire control of the national policy, and almost without exception, the possession of the highest executive office of the Union. Always united in the purpose of regulating the affairs of the whole Union by the standard of the slave-holding interest, their disproportionate numbers in the electoral college have enabled them, in ten out of twelve quadrennial elections, to confer the Chief Magistracy upon one of their own citizens. At this moment, the President of the United States, the President of the Senate, the Speaker of the House of Representatives, and the Chief Justice of the United States, are all citizens of the favored portion of the united republic. . . .
>
> It is worthy of observation that this slave representation is always used to protect and extend slave power; and in this way, the slaves themselves are made to vote for slavery: they are compelled to furnish halters to hang their posterity . . .

Slave owners or their allies controlled all the branches of federal government throughout most of the pre–Civil War period.

Yes, writes Ira Berlin, one of the leading historians of slavery,

> the founders were slaveholders, and substantial ones. At one time or another, each condemned slavery as evil and recognized the contradiction of his own stated beliefs to the realities of chattel bondage. None, however, had the courage to translate his intellectual commitment to universal freedom into public policy or even individual action. Only Washington freed his slaves, although Jefferson freed a few favorites, most of whom were kin. Instead, they "celebrated freedom while stealing the substance of life from the people they 'owned.'"

Just as the new nation was getting started, the New Englander Eli Whitney traveled south to take on a new job. . . .

WITH THE CIVIL WAR OVER, A GROUP OF EMANCIPATED SLAVES GATHERS ON A RIVERFRONT DOCK IN VIRGINIA.

Mr. Whitney Can Make Anything

Eli Whitney was born on December 5, 1765, and spent his youth on the family farm near Westboro, Massachusetts. He was only eleven when the fighting at nearby Lexington and Concord ignited the Revolutionary War. His mother, long sick, died when he was twelve, and when his father remarried two years later, a stepmother and two stepsisters became his family.

With war came hard times. Prices shot up for nearly everything. Even nails, usually imported, cost too much for the building and repair work always needed around farms. Early on, Eli showed a native skill for mechanical work. (He had made a playable violin at twelve.)

Eli had "an instinctive understanding of mechanisms," said the historian Mitchell Wilson. "It was a medium in which he could improve and create in exactly the same way that a poet handles words or a painter uses color." He talked his father into letting him install a forge. On a machine he made at home, he began producing nails cheaply enough for everyone to buy. Soon the demand was so great that young Eli, without asking his dad's permission, scoured the countryside to find a workman who could help out. The two visited many workshops, searching out

PORTRAIT OF THE INVENTOR ELI WHITNEY WHEN HE WAS ABOUT FIFTY

tools and techniques that would improve how things were made or repaired.

When the war was won and peace arrived, foreign products, including nails as cheap as the ones Eli made, again came flowing into the United States. So he shifted production and became the sole American producer of hat pins for women.

As he approached eighteen, Whitney decided he had to go to college. But there was no money to pay tuition. In the hope of earning enough, he found a job teaching school. In those early times, all you needed to become a teacher was to win the town's approval. He taught the country school for five winters, reading as widely as possible to keep one jump ahead of the youngsters. Summers, to prepare himself for Yale's entrance exams, he took classes at an academy nearby.

Then twenty-three, Whitney thought he was ready for college. It was a very late age to become a freshman. In the coastal towns of Massachusetts, many twenty-year-olds were already the captains of whaling vessels that sailed around the world for years hunting their prey. Whitney's stepmother thought attending college was a bad idea, but his father promised to help with the tuition, and in 1789, he set off for Yale College in New Haven, Connecticut.

Whitney's years at Yale did a lot for him. The intellectual excitement of the academic world, the stimulation of classmates and professors, and the new friends he made all helped to bring his natural talents to fruition. Soon, and later in life as well, influential Yale men would be able to help him in times of need.

At college, he took on odd jobs to help pay his expenses, but he still had to turn to his father: "I have succeeded very well in my studies," he wrote, "and meet with no other difficulties but the want of money, which indeed is very great."

In September 1792, Whitney graduated from Yale. Without a dime, however, and now twenty-seven, what would he do to support himself?

He decided to study law. At that time, you could, by reading the legal literature intensively, prepare yourself for the bar. Meanwhile, to secure some income, he agreed to tutor the children of a South Carolina planter named Major Dupont.

He sailed south on a coastal packet boat that took a few passengers. Among them was the widow of General Nathanael Greene, a Revolutionary War veteran. The Greenes had settled in Savannah, Georgia, after the war. Mrs. Greene, with her children and her estate manager, Phineas Miller, were returning home after a visit to northern friends.

When Whitney arrived in South Carolina, he found that the salary promised him for tutoring would be halved. Disgusted, he refused the job and decided to quit teaching, too. Mrs. Greene invited him to come with her to Mulberrry Grove, the plantation given to the general by the state of Georgia. You can read law, she said, and help Phineas Miller, my plantation manager. Miller was a Yale graduate, too, a few years older than Eli. It seemed an agreeable prospect, and Whitney decided to stay.

Two weeks after his arrival, Whitney had invented the cotton gin.

How did that near-miracle happen?

Shortly after Whitney settled in at Mrs. Greene's, some neighbors dropped in and the talk quickly turned to how bad times were. They couldn't make money from their crop. The only variety of cotton that would grow in their area was the awful green-seed kind. It took ten hours of handwork by their slaves to separate one pound of lint from three pounds of the small, tough seed. That was terribly time-consuming and expensive. Nearly as many slaves were needed to separate the seed and boll from the cotton as were needed to harvest the crop itself.

One planter grumbled that the green-seed cotton wasn't much better than a useless weed. If only some kind of machine could be invented to do the work!

PICKING COTTON IN THE MISSISSIPPI DELTA REGION

"Gentlemen," spoke up Mrs. Greene, "apply to my friend Mr. Whitney. He can make anything."

Urged on by Mrs. Greene and Phineas Miller, Whitney went out to watch the slaves cleaning the cotton. He studied their hand movements. He saw that one hand held the seed while the other hand plucked out the short strands of lint. And he saw almost at once how to create a machine to do exactly what the slave was doing, but better and ever so much faster.

A gin for black-seed cotton existed. The fibers passed between two rollers, and the smooth seed dropped out. But the fuzzier green seed would pass right on through, mashed into the fibers. That gin didn't work very well.

The machine Whitney devised replaced hand-holding the seed with a kind of sieve of wires stretched lengthwise. To do the work of the fingers that pulled out the lint, he had a drum rotate past the sieve, almost

AN ARTIST'S DEPICTION OF THE YOUNG ELI WHITNEY IN HIS WORKSHOP

touching it. On the surface of the drum, he placed fine hook-shaped wires that caught at the lint. The restraining wires of the sieve held the seed back while the lint was pulled away. A rotating brush which turned four times as fast as the hook-covered drum cleaned the lint off the hooks.

Sound simple? It was. There was no need to make it more complicated.

When his new machine was ready, Whitney demonstrated that first model, calling in a few neighboring planters to see how it operated. They were staggered. In one hour, it produced the full day's work of several slaves. "The machine could be turned by water, horses, or in any other way as is most convenient," its inventor said. But its very simplicity would cause Whitney trouble and pain for years to come.

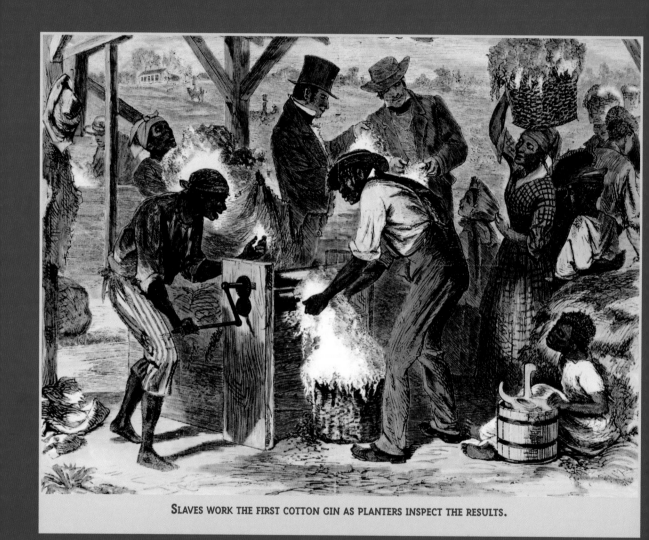

SLAVES WORK THE FIRST COTTON GIN AS PLANTERS INSPECT THE RESULTS.

Inventions: Good for Whom?

The men who witnessed Whitney's demonstration couldn't wait to use it on their own plantations. He'd make more cotton gins, Whitney promised, and get a patent to protect his invention. The local planters immediately had their slaves plant great swatches of land with green-seed cotton.

Eli was delighted with the prospect of great riches. He formed a partnership with Miller, his fellow Yale man, to exploit the invention. Two mistakes blighted their chances. They failed to guard the model from predators. The workshop at Mrs. Greene's was broken into, and the model of the gin carefully examined and sketched. It's called piracy—theft of what someone else owns.

The other mistake the partners made was to let hunger for quick profits overrule their better judgment. They could have made and sold gins directly to customers or sold the right to manufacture them. Instead, they decided to place their gins at key points in the cotton regions and charge two-thirds of a pound of cotton for every pound

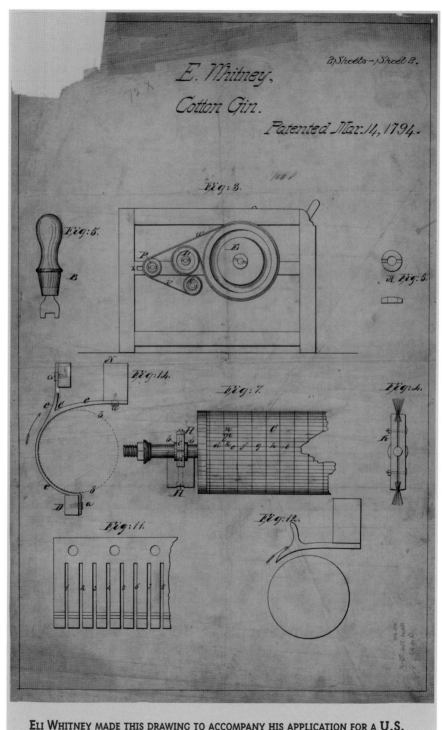

returned to the planter. They couldn't have known at the time that cotton planting would spread like an epidemic and that their proposal would have netted them millions of dollars a year. Once cotton planting took off, the amount they charged seemed like an outrageously high rate of profit to prospective customers.

Instead of making such a deal, planters had their own mechanics copy the simple structure of the gin. Soon, the whole South was busily and happily ginning huge masses of cotton.

Whitney went up to Philadelphia to apply for a patent from Secretary of State Thomas Jefferson, in whose department the patent office was placed. Meanwhile, Phineas Miller tried to locate the best places in the cotton states to install the new gins. While awaiting a decision on the patent, Whitney opened a shop in New Haven to build gins. Bad luck struck when a fire left the new cotton-gin factory "all in ashes! My shop, all my tools, material and work equal to twenty finished cotton machines all gone. . . ."

To manufacture a gun, or almost anything else in Whitney's time, meant to make it by hand. A workman hammered, filed, carved out, and fitted a gun. And then he started over and in the same slow and tedious way made another and another. After his invention of the cotton gin, Whitney seized upon the idea that if each part of the thing being made was a standard size and shape, the complete thing could be assembled at any time later.

He worked to apply the idea to the manufacture of muskets for the federal government, under a contract given him in 1798. It called for producing ten thousand muskets in two years for $134,000. It took him ten years to finish the job, with all kinds of headaches along the way.

What counts is the key idea: the mass production of things. "One of my primary objects," he said, "is to form the tools so that the tools themselves shall fashion the work." The idea was not original. A French mechanic some twenty years earlier had come up with it and discussed it with Thomas Jefferson.

But whoever deserves the credit, the concept, used more and more, would change the entire method of manufacturing. Standardization of parts, making them interchangeable, came about through the creation of machines of accurate caliber. An entire family of machine tools evolved, including milling, grinding, and boring machines.

Whitney worked out and built all the machinery he would need for this revolutionary method of production. His gun factory was the first to use mass-production techniques.

WOMEN AT WORK AT A COTTON WAREHOUSE IN CHARLESTON, SOUTH CAROLINA

The fire didn't matter to the planters. They were already ginning cotton with gins they had built without Whitney.

Whitney got his patent, but it didn't stop others from infringing upon it. He cried out that his invention was making southerners rich while he wasn't earning a penny. It was too late and too costly to sue individual planters for the wrong done him. Instead, he and Miller pleaded for the cotton states to redress the wrong. Perhaps pay them a flat fee in return for which the cotton gin would be public property in the state? In 1801, South Carolina agreed, and authorized a $50,000 payment for which a down payment of $20,000 was made, and never any more. Other states too paid something. It was small comfort, because the partners' legal and other expenses were barely covered. In 1803, the states repudiated their agreements and sued Whitney for all the money paid him and Miller. That year alone the planters earned $10 million from their cotton crop. And the price of slaves had doubled. The patent expired in 1807, and Whitney's hopes of success and recognition died with it.

Some questions were raised about whether Whitney had conceived the plan for the cotton gin all by himself. As with many inventions, priority or originality was contested. Was someone else's idea involved? It was said that Mrs. Greene had added a significant element when she suggested that Whitney use a hook-shaped wire on the drum. Others claimed that the gin was a common device of the time. Still others asserted that it was slaves on the plantation who gave Whitney the basic idea. He simply mechanized it.

Because documentary evidence is missing, we have no way of knowing whether any of these claims are valid.

Within a few years Whitney's invention had changed the South's economy. It took comparatively little capital to raise cotton, so many farmers switched from tobacco, indigo, or rice. Production rose, more land was seeded with cotton, and more black labor was bought to raise the crops.

Soon cotton was America's leading crop and the chief export commodity. The textile mills of the northern states and of Britain bought at good prices all the cotton the planters could provide. In England alone, according to an article in the *London Economist*, "the lives of nearly two millions of our countrymen are dependent upon the cotton crops of America; their destiny may be said, without any kind of hyperbole, to hang upon a thread. Should any dire calamity befall the land of cotton, a thousand of our merchant ships would rot idly in dock; ten thousand mills must stop their busy looms; two thousand mouths would starve, for lack of food to feed them." But there were no signs of calamity; instead, confidence in the profitable crop increased.

Another of the many signs of the great importance of the invention can be seen in its effects on the New Orleans economy in slavery days. Investors provided $754,000—an enormous sum then—for the building of the New Orleans Cotton Press. It had a capacity of producing 150,000 bales and warehouses for storing 25,000 bales. (On average, a bale is a compressed bundle weighing between 400 and 500 pounds [180 and 230 kilograms].)

With the arrival of prosperity, large-scale cultivation by the bigger and more efficient planters pushed out the smaller, undercapitalized farmers.

Cotton became king. What crowned it king was Whitney's invention. We hear the word *invention* and almost automatically think "progress." Yes, progress in the sense of improved technology, a better way of doing things or making things. Still, there are questions about "progress." Did the invention improve life for everyone? Or did it benefit some while doing great harm to others?

The cotton gin was of no material benefit to its inventor. Whitney, as we've seen, didn't profit financially from his ingenuity, only in reputation.

It was the cotton planter who profited enormously by Whitney's gin. Not only in money, but in prestige and in power.

While the Yankee's cotton gin made the cotton planters rich, it also made millions of black men, women, and children slaves, and great numbers of workers in cotton mills overworked and underpaid, the victims of exploitation.

A Mississippi steamboat, loaded with bales of cotton, docks at Baton Rouge, Louisiana.

Labor in the Mills

It is no secret that workers in the nineteenth century were badly treated. Industrialists were ruthless in placing their concern for low costs and high profits above all else. They drove their employees to produce more and more without offering any reward in the form of higher pay or better hours.

But "badly as our workingmen may be treated," said one labor newspaper, "the condition of females who are obliged to work for a living is far worse."

Women were used to hard work. At home, they often toiled for more than twelve hours a day, doing such chores as cleaning, cooking, and sewing, all while raising children. Their labor was heavy, unending—and unpaid. When employers sent agents into rural districts to recruit farm girls, thousands responded. The picture of life in the mill towns—high wages, leisure hours, silk dresses—was a lure they could not resist.

But the beauties of factory life proved a myth. Women worked fourteen to sixteen hours a day for a wage of about $1.50 a week.

The Civil War brought even more women to the mills. After the war, the number increased, partly because, as one government official put it,

GHOSTLY FACES PEER DOWN THE LONG ROWS. WOMEN WERE A FIXTURE IN
COTTON MILLS AROUND THE WORLD, SUCH AS THIS ONE IN MALAGA, SPAIN.
THEY HELPED DRIVE THE SPREAD OF THE INDUSTRY.

women did more and better work than many men who were paid twice as much. Thus their employment was greatly desired.

The women who did piece work at home, usually sewing garments by hand, were no better off. With the invention of the sewing machine, the rates for piece work plummeted. The women had to supply their own thread, at ten cents per spool. In 1870, women were getting six cents for each shirt they made. A survey revealed that in New York, 7,000 working women could afford to live only in cellars, and 20,000 were near starvation.

In the cotton mills of New England and the South, tens of thousands of women labored long and dreary hours. Marie Van Vorst, one of the early investigators of labor conditions, entered some South Carolina mills to see what the women did. "Do you like the mills?" she asked. Without exception the answer was "No" or even "I hate them." According to Van Vorst, the reasons were many:

A WOMAN DRAWS THREADS INTO A POWER LOOM IN A DALLAS, TEXAS, MILL. THE WORK REQUIRED CONCENTRATION, DEXTERITY, AND ENDURANCE. THE LONG HOURS CREPT BY, ALL TO THE ATTENDANT HUM OF THE MACHINERY CONSTANTLY IN MOTION.

"Spooling" is hard on the left arm and the side. Heart disease is a frequent complaint amongst the older spoolers. The cotton comes from the spinning-room to the spool-room, and as the girl stands before her

UNDER THE WATCHFUL EYE OF HER SUPERVISOR, A YOUNG GIRL PAUSES FROM HER WORK IN A MILL.

IT WAS THE PLIGHT OF CHILDREN SUCH AS THIS YOUNG SOUTH CAROLINA GIRL THAT MARIE VAN VORST WOULD BRING TO A NATIONAL AUDIENCE THROUGH HER VARIOUS EXPOSÉS. CHILD LABOR REFORMS WOULD SLOWLY COME.

"side," as it is called, she sees on a raised ledge, whirling in rapid vibration, some one hundred huge spools full of yarn, whilst below her, each in its little case, lies a second bobbin of yarn wound like a distaff [a long rod used for spooling thread or yarn].

Her task controls machinery in constant motion, then never stops except in case of accident. With one finger on her right hand she detaches the yarn from the distaff that lies inert in the little iron rut before her. With her left hand she seizes the revolving circle of the large spool's top in front of her, holding this spool steady, overcoming the machinery for the moment not as strong as her grasp. This demands a certain effort. Still controlling the agitated spool with her left hand, she detaches the end of yarn with the same hand from the spool, and by means of a patent knotter harnessed around her palm she joins together the two loosened ends, one from the little distaff and one from this large spool, so that the two objects are set whirling in unison and the spool receives all yarn from the distaff. Up and down this line the spooler must walk all day long, replenishing the iron grooves with fresh yarn and reknitting broken strands. . . .

The air of the room is white with cotton, although the spool-room is perhaps the freest. These little particles are breathed into the nose, drawn into the lungs. Lung disease and pneumonia—consumption—are the constant, never-absent scourge of the mill village. The girls expectorate [cough and spit] to such an extent that the floor is nauseous with it.

Even the method of payment was degrading, the reporter found:

Some of the hands never touch their money from month's end to month's end. Once in two weeks is payday. A woman had then worked 122 hours. The corporation furnishes her house. There is the rent to be paid; there are also the corporation stores from which she has been getting her food and coal and what gewgaws [trinkets] the cheap stuff on sale may tempt her to purchase. There is a book of coupons issued by the mill owners which are as good as gold. It is good at the stores, good for the rent, and her time is served out in pay for this representative currency. This is of course not obligatory, but many of the operatives avail themselves or bind themselves by it. When the people are ill, they are docked for wages. When, for indisposition or fatigue, they knock a day off, there is a man hired especially for this purpose, who rides from house to house to house to find out what is the matter with them, to urge them to rise, and if they are not literally too sick to move, they are hounded out of their beds and back to their looms.

A SHIFT IN THE MILL COULD BE DANGEROUS WORK. SOME EMPLOYERS LIKED HIRING CHILDREN BECAUSE THEIR HANDS COULD SLIP INTO SMALL OPENINGS IN THE MACHINES TO MAKE REPAIRS OR REPLACE PARTS. INDUSTRIAL ACCIDENTS WERE NOT COMMON, BUT A CONSTANT THREAT. THESE BOYS HAVE CLIMBED ON TOP OF THE MACHINE AS THEY REPLACE BOBBINS, REPAIRING BROKEN THREADS IN THE BIBB COTTON MILL IN MACON, GEORGIA.

INSTEAD OF LEAVING FOR SCHOOL, THESE MISSISSIPPI GIRLS ARE HEADING OFF TO A LONG DAY AT THE COTTON MILL. THEY WALK PAST THE ROW OF FRAME HOUSES OCCUPIED BY OTHER MILL WORKERS IN THE TOWN.

Children were even worse off in the textile industry. In 1883, a U.S. Senate committee, investigating "the relations between labor and capital," looked into child labor in the cotton mills. Thomas L. Livermore, manager of the Amoskeag mills in Manchester, New Hampshire, was one of the people who offered testimony:

Question: *Won't you please tell us your experience with the question of child labor; how it is, and to what extent it exists here; why it exists, and whether, as it is actually existing here, it is a hardship on a child or on a parent; or whether there is any evil in that direction that should be remedied?*

Answer: These is a certain class of labor in the mills which, to put it in very common phrase, consists mainly in running about the floor—where there is not as much muscular exercise required as a child would put forth in play, and a child can do it about as well as a grown person can do it—not quite as much of it, but somewhere near it, and with proper supervision of older people, the child serves the purpose. That has led to the employment of children in the mills, I think. . . .

Now, I think that when it is provided that a child shall go to school as long as it is profitable for a workman's child (who has got to be a workingman himself) to go to school, the limit has been reached at which labor in the mills should be forbidden. There is such a thing as too much education for working people sometimes. I do not mean to say by that that I discourage education to any person on earth, or that I think that with good sense any amount of education can hurt any one, but I have seen cases where young people were spoiled for labor by being educated to a little too much refinement.

But there was little worry that the children might become a little too refined to be useful in the mills. Twenty years later, in 1903, Marie Van Vorst found children five, six, and seven years old amid the looms of South Carolina's mills.

Through the looms I catch sight of Upton's, my landlord's little child. She is seven; so small that they have a box for her to stand upon. She is a pretty, frail, little thing, a spooler—"a good spooler, tew!" Through the frames on the other side I can only see her fingers as they clutch at the flying spools; her head is not high enough, even with the box, to be visible. Her hands are fairy hands, fine-bones, well-made, only they are so thin and dirty, and nails—claws: she would do well to have them cut. A nail can be torn from the finger, is torn from the finger frequently, by this flying spool. I go over to Upton's little girl. Her spindles are not thinner nor her spools whiter.

"How old are you?"

"Ten."

She looks six. It is impossible to know if what she says is true. The children are commanded both by parents and bosses to advance their ages when asked.

"Tired?"

She nods, without stopping. She is a "remarkable fine hand." She makes 40 cents a day. See the value of this labor to the manufacturer, cheap, yet skilled; to the parent it represents $2.40 per week. . . .

Marie Van Vorst next turned her attention to the activities of a young boy:

Besides being spinners and spoolers, and weavers even, the children sweep the cotton-strewed floors. Scarcely has the miserable little object, ragged and odorous, passed me with his long broom, which he drags half-heartedly along, than the space he has swept up is cotton-strewn again. It settles with discouraging rapidity; it has also settled on the child's hair and clothes, and his eyelashes, and this atmosphere he breathes and fairly eats, until his lungs become diseased. . . .

Here is a little child, not more than five years old. The land is a hot enough country, we will concede, but not a. . . . South Sea Island! She has on one garment if a tattered sacking dress can so be termed.

A TEN-YEAR-OLD GIRL DONS HER FINEST OUTFIT, COMPLETE WITH A HAT, TO WORK IN THE SPINNING ROOM OF A ROANOKE, VIRGINIA, MILL.

Her bones are nearly through her skin, but her stomach is an unhealthy pouch, abnormal. She has dropsy, she works in a new mill—in one of the largest mills in South Carolina. Here is a slender little boy—a birch rod (good old simile) is not more slender, but the birch has the advantage: it is elastic—it bends, has youth in it. This boy looks ninety. He is a dwarf; twelve years old, he appears seven, no more. . . . He sweeps the cotton and lint from the mill aisles from 6 P.M. to 6 A.M. without a break in the night's routine. He stops of his own accord, however, to cough and expectorate—he has advanced tuberculosis.

At night the shanties receive us. On a pine board is spread our food—can you call it nourishment? The hominy and molasses is the best part; salt pork and ham are strong victuals.

It is eight o'clock when the children reach their homes—later if the mill work is behind-hand and they are kept over hours. They are usually beyond speech. They fall asleep at the table, on the stairs; they are carried to bed and there laid down as they are, unwashed, undressed; and the inanimate bundles of rags so lie until the mill summons them with its imperious cry before sunrise, while they are still in stupid sleep.

That was a southern mill. But it was not much different in the North, as is evident from the reports of John Spargo, who studied New England's factories in 1906. Around 1900, it was estimated that at least 80,000 children, most of them girls, were employed in the country's textile mills.

Witnesses have described the health effects of long hours of hard work. Some kinds of work were more harmful than others. In general, the workplace was often unsanitary, badly ventilated, or put children in the presence of workers who were a bad social and moral influence.

There was still another way those children tending the mill machines were harmed. Their education ceased when they began factory work. With no future chance to learn, the doors to a better life were shut. Their labor was a dreary road ended only by injury, sickness, or age.

Adult Cares

How did those children feel? What did they think about their lives? This poem, published in a magazine in 1832, attempted to capture the working life as seen through the eyes of a child:

I often think how once we used in summer
        fields to play,
And run about and breathe the air that made us
        glad and gay;
We used to gather buttercups and chase the
        butterfly;
I loved to feel the light breeze lift my hair as it
        went by!

Do you still play in those bright fields? And are
        the flowers still there?
There are no fields where I live now—no
        flowers any where!
But day by day I go and turn a dull and
        tedious wheel;
You cannot think how sad, and tired, and faint
        I often feel.

I hurry home to snatch the meal my mother can
        supply,
Then back I hasten to the task—that not to
        hate I try.
At night my mother kisses me, when she has
        combed my hair,
And laid me in my little bed, but—I'm not
        happy there:

I dream about the factory, the fines that on us
        wait—
I start and ask my father if I have not lain
        too late?
And once I heard him sob and say—"Oh better
        were a grave,
Than such a life as this for thee, thou little sinless slave!"

I wonder if I ever shall obtain a holiday?
Oh, if I do, I'll go to you and spend it all in
        play!
And then I'd bring some flowers home, if you
        will give me some,
And at my work I'll think of them and
        holidays to come!

Picking Cotton

When cotton, once a minor crop in the South, became its number one product, it led to the resurgence of the slave trade. Tens of thousands of slaves were imported to harvest bumper crops of cotton. In 1793, the year of Whitney's invention, 10,000 bales of cotton were produced. In 1825, more than 500,000 bales. In 1860, 5 million bales. Three-fourths of the world's cotton was produced in the South. More and more traders got into the business of buying and selling slaves. Little sympathy was extended to the Africans. With money to be made, black people were simply commodities that were used to make others rich.

Although the cotton gin greatly diminished the time needed to process cotton, it did nothing to lessen the time needed to harvest it. The planters were eager for more and more workers to pick and bale the cotton. And that led to a great rise in the slave population.

By the early 1800s, more than 700,000 slaves lived in the South. They made up a third of the region's population. By 1860, the eve of the Civil War, the Southern states had about 4 million slaves. In two states—South Carolina and Mississippi—they actually outnumbered the whites.

A SLAVE FAMILY IN SOUTH CAROLINA, PHOTOGRAPHED IN 1862

A WHITE OVERSEER ON A TEXAS PLANTATION SUPERVISES SLAVES AT WORK.

The large plantation was a business. Its purpose was to produce a crop, cotton in this case, for sale on the market. Operating as a capitalist, the planter viewed his slaves as tools of production to be used for the greatest profit. In 1850, on plantations that grew the South's five great staple crops—cotton, tobacco, sugar, rice, and hemp—by far the greatest number of slaves (1,815,000) lived on cotton plantations. That number was three times greater than the number of slaves raising and harvesting all other crops combined.

Most white Southerners were not planters. Three out of four had no ties to slavery, either through personal ownership or family. They were mostly small farmers working their own fields, with only family help.

To be a planter with any standing, you had to own at least twenty slaves. Only 12 percent of the planters were in that class. About half the slaveholders owned fewer than five slaves. The majority of slaves was in the hands of the large planters. Around 3,000 of the planters owned 100 slaves or more.

What was it like to be a slave on a cotton plantation? For evidence, we have the report of Solomon Northup. Born a free black, he was kidnapped in Washington, D.C., in 1841 and forced into slavery on a cotton plantation near the Red River in Louisiana. It was twelve years before he managed to escape to freedom. He then wrote the story of how he lived through those long years. Here he describes picking cotton:

The ground is prepared by throwing up beds or ridges, with the plough—back-furrowing, it is called. Oxen and mules, the latter almost exclusively, are used in ploughing. The women as frequently as the men perform this labor, feeding, currying, and taking care of the teams, and in all respects doing the field and stable work, precisely as do the ploughboys of the North.

The beds, or ridges, are six feet wide, that is, from water furrow to water furrow. A plough drawn by one mule is then run along the top of the ridge or center of the bed, making the drill, into which a girl usually drops the seed, which she carries in a bag hung round her neck. Behind her comes a mule and harrow, covering up the seed, so that two mules, three slaves, a plough and harrow, are employed in planting a row of cotton.

This is done in the months of March and April. Corn is planted in February. When there are no cold rains, the cotton usually makes its appearance in a week. In the course of eight or ten days afterwards the first hoeing is commenced. This is performed in part, also, by the aid of the plough and mule. The plough passes as near as possible to the cotton on both sides, throwing the furrow from it. Slaves follow with their hoes, cutting up the grass and cotton, leaving hills two feet and a half apart. This is called scraping cotton.

In two weeks more commences the second hoeing. This time the furrow is thrown towards the cotton. Only one stalk, the largest, is now left standing in each hill. In another fortnight it is hoed the third time, throwing the furrow towards the cotton in the same manner as before, and killing all the grass between the rows.

A PLANTATION MANAGER WEIGHS THE COTTON BROUGHT IN BY THE FIELD HANDS.

CHILDREN WORK ALONGSIDE THEIR PARENTS IN THE COTTON FIELDS NEAR SAVANNAH, GEORGIA.

About the first of July, when it is a foot high or thereabouts, it is hoed the fourth and last time. Now the whole space between the rows is ploughed, leaving a deep water furrow in the center. During all these hoeings the overseer or driver follows the slaves on horseback with a whip, such as has been described. The fastest hoer takes the lead row. He is usually about a rod in advance of his companions. If one of them passes him, he is whipped. If one falls behind or is a moment idle, he is whipped. In fact, the lash is flying from morning until night, the whole day long. The hoeing season thus continues from April until July, a field having no sooner been finished once, than it is commenced again.

In the latter part of August begins the cotton picking season. At this time each slave is presented with a sack. A strap is fastened to it, which goes over the neck, holding the mouth of the sack breast high, while the bottom reaches nearly to the ground. Each one is also presented with a large basket that will hold about two barrels. This is to put the cotton in when the sack is filled. The baskets are carried to the field and placed at the beginning of the rows.

Northup tells how a beginning picker is treated:

When a new hand, one unaccustomed to the business, is sent for the first time into the field, he is whipped up smartly, and made for that day to pick as fast as he can possibly. At night it is weighed, so that his capability in cotton picking is known. He must bring in the same weight each night following. If it falls short, it is considered evidence that he has been laggard, and a greater or less number of lashes is the penalty.

An ordinary day's work is two hundred pounds [ninety kilograms]. A slave who is accustomed to picking is punished if he or she brings in a less quantity than that. There is a great difference among them as regards this kind of labor. Some of them seem to have a natural knack, or quickness, which enables them to pick with great celerity, and with

A COTTON
FIELD IN GEORGIA

69

both hands, while others, with whatever practice or industry, are utterly unable to come up to the ordinary standard. Such hands are taken from the cotton field and employed in other business. . . .

From "day clear" to "first dark," the slaves stayed in the fields.

The hands are required to be in the cotton field as soon as it is light in the morning, and, with the exception of ten or fifteen minutes, which is given them at noon to swallow their allowance of cold bacon, they are not permitted to be a moment idle until it is too dark to see and when the moon is full they often times labor till the middle of the night. They do not dare to stop even at dinner time, nor return to the quarters, however late it be, until the order to halt is given by the driver.

The day's work over in the field, the baskets are "toted," or in other words, carried to the gin-house, where the cotton is weighed. No matter how fatigued and weary he may be—no matter how much he longs for sleep and rest—a slave never approaches the gin-house with his basket of cotton but with fear. If it falls short in weight—if he has not performed the full task appointed him, he knows that he must suffer. And if he has exceeded it by ten or twenty pounds, in all probability his master will measure the next day's task accordingly.

So, whether he has too little or too much, his approach to the gin-house is always with fear and trembling. Most frequently they have too little, and therefore it is they are not anxious to leave the field. After weighing, follow the whippings; and then the baskets are carried to the cotton house, and their contents stored away like hay, all hands being sent in to tramp it down. If the cotton is not dry, instead of taking it to the gin-house at once, it is laid upon platforms, two feet [sixty centimeters] high, and some three times as wide, covered with boards or plank, with narrow walks running between them.

But that by no means ended the day, said Northup:

Each one must then attend to his respective chores. One feeds the mules; another the swine—another cuts the wood, and so forth; besides, the packing is all done by candle light. Finally, at a late hour, they reach the quarters, sleepy and overcome with the long day's toil. Then a fire must be kindled in the cabin, the corn ground in a small hand-mill, and supper, and dinner for the next day in the field, prepared. All that is allowed them is corn and bacon, which is given out at the corncrib and smoke-house every Sunday morning. Each one receives, as his weekly allowance, three and a half pounds of bacon, and corn enough to make a peck of meal. That is all—no tea, coffee, sugar, and with the exception of a very scanty sprinkling now and then, no salt. . . .

And their housing? Northup tells what that was like:

The softest couches in the world are not to be found in the log mansion of the slave. The one whereon I reclined, year after year, was a plank twelve inches [thirty cm] wide and ten feet [three meters] long. My pillow was a stick of wood. The bedding was a coarse blanket, and not a rag or shred beside. Moss might be used, were it not that it directly breeds a swarm of fleas.

The cabin is constructed of logs, without floor or window. The latter is altogether unnecessary, the crevices between the logs admitting sufficient light. In stormy weather the rain drives through them, rendering it comfortless and extremely disagreeable. The rude door hangs on great wooden hinges. In one end is constructed an awkward fire-place.

An hour before day light the horn is blown. Then the slaves arouse, prepare their breakfast, fill a gourd with water, in another deposit their dinner of cold bacon and corn cake, and hurry to the field again. It is

A SHIRTLESS SLAVE STANDS WITH HIS HANDS TIED TO A WHIPPING POST AS A WHITE MAN PREPARES TO LASH HIM.

an offence invariably followed by a flogging, to be found at the quarters after daybreak. Then the fears and labors of another day begin; and until its close there is no such thing as rest. He fears he will be caught lagging through the day; he fears to approach the gin-house with his basket-load of cotton at night; he fears, when he lies down, that he will oversleep himself in the morning. Such is a true, faithful, unexaggerated picture and description of the slave's daily life, during the time of cotton-picking, on the shore of Bayou Boeuf.

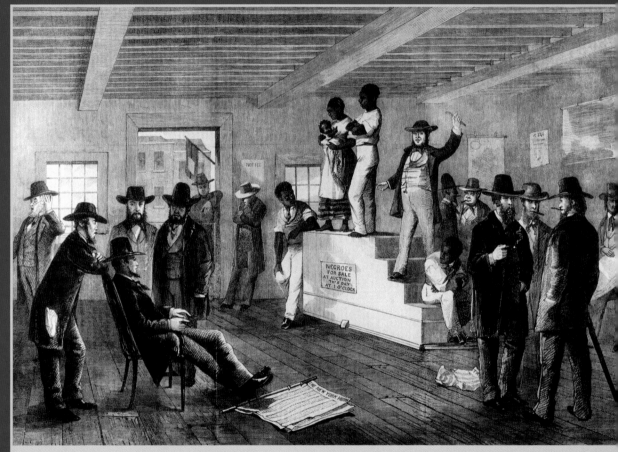

A FAMILY OF SLAVES IS ON THE AUCTION BLOCK IN RICHMOND, VIRGINIA.

On the Auction Block

As cotton cultivation spread south and west, slaves from the upper South were sold to planters in the richer cotton regions. The auction block—a routine business market for the whites—was a dreaded place for the slaves. Men, women, and children who were going to be sold were held in slave pens, which existed in all the larger cities of the South.

Solomon Northup, who experienced several of them, described the slave pen in New Orleans, owned by Theophilus Freeman (an ironic name!). It stood opposite the St. Charles Hotel. Three tiers of rooms housed the slaves. There were two entrances: one for the buyers, the other for the slaves. He tells what an auction was like at Freeman's place:

> In the first place we were required to wash thoroughly, and those with beards to shave. We were then furnished with a new suit each, cheap, but clean. The men had hat, coat, shirt, pants and shoes; the women frocks of calico, and handkerchief to bind about their heads. We were now conducted into a large room in the front part of the building to which the yard was attached, in order to be properly trained, before

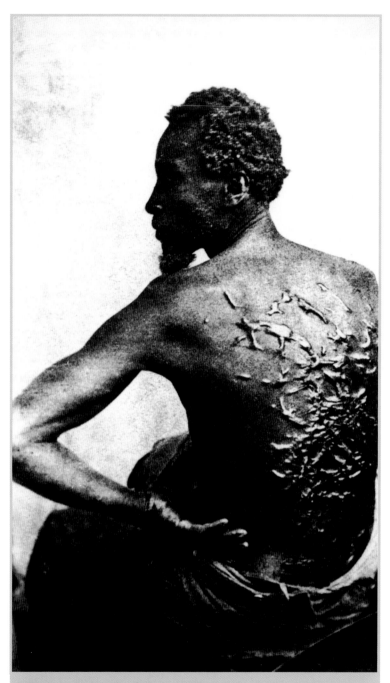

GORDON, A RUNAWAY SLAVE FROM MISSISSIPPI, ESCAPED THROUGH
UNION LINES DURING THE CIVIL WAR. HIS BACK IS BARED TO SHOW THE
SCARS OF REPEATED WHIPPINGS. THIS PHOTOGRAPH WAS REPRODUCED IN
HARPER'S WEEKLY IN 1863.

the admission of customers. The men were arranged on one side of the room, the women at the other. The tallest was placed at the head of the row, then the next tallest, and so on in the order of their respective heights. Emily was at the foot of the line of women. Freeman [owner of the slave pen] charged us to remember our places; exhorted us to appear smart and lively,—sometimes threatening, and again, holding out various inducements. During the day he exercised us in the art of "looking smart," and of moving to our places with exact precision.

After being fed, in the afternoon, we were again paraded and made to dance. Bob, a colored boy, who had some time belonged to Freeman, played on the violin. . . .

Next day many customers called to examine Freeman's "new lot." The latter gentleman was very loquacious, dwelling at much length upon our several good points and qualities. He would make us hold up our heads, walk briskly back and forth, while customers would feel of our hands and arms and bodies, turn us about, ask us what we could do, make us open our mouths and show our teeth, precisely as a jockey examines a horse which he is about to barter for or purchase. Sometimes a man or woman was taken back to the small house in the yard, stripped, and inspected more minutely. Scars upon a slave's back were considered evidence of a rebellious or unruly spirit, and hurt his sale.

During the day a number of sales were made. David and Caroline were purchased together by a Natchez planter. They left us grinning broadly, and in a most happy state of mind, caused by the fact of their not being separated.

One buyer showed interest in a ten-year-old boy named Randall, who was with his mother, Eliza, and half-sister Emily:

The little fellow was made to jump, and run across the floor, and perform many other feats, exhibiting his activity and condition. All the time the trade was going on, Eliza was crying aloud, and wringing her hands. She besought the man not to buy him, unless he also bought herself and Emily. She promised, in that case, to be the most faithful slave that ever lived. The man answered that he could not afford it, and then Eliza burst into a paroxysm of grief, weeping plaintively. Freeman turned round to her, savagely, with his whip in his uplifted hand, ordering her to stop her noise, or he would flog her. He would not have such work—such sniveling; and unless she ceased that minute, he would take her to the yard and give her a hundred lashes. Yes, he would take the nonsense out of her pretty quick—if he didn't might he be d——d. Eliza shrunk before him, and tried to wipe away her tears, but it was all in vain. She wanted to be with her children, she said, the little time she had to live.

All the frowns and threats of Freeman could not wholly silence the afflicted mother. She kept on begging and beseeching them, most piteously, not to separate the three. Over and over again she told them how she would labor day and night, to the last moment of her life; if he would only buy them all together. But it was of no avail; the man could not afford it. The bargain was agreed upon, and Randall must go alone. Then Eliza ran to him; embraced him passionately; kissed him again and again; told him to remember her—all the while her tears falling in the boy's face like rain.

Freeman d——d her, calling her a blubbering, bawling wench, and ordered her to go to her place, and behave herself, and be some-body. He swore he wouldn't stand such stuff but a little longer. He would soon give her something to cry about, if she was not mighty careful, and that she might depend upon.

The planter from Baton Rouge, with his new purchase, was ready to depart.

"Don't cry, mama. I will be a good boy. Don't cry," said Randall, looking back, as they passed out of the door.

A heartrending personal example of what slavery could do to shatter black families came from Edward Ball in 1999. In his book, *Slaves in the*

AN 1852 PAINTING DEPICTS A SLAVE AUCTION AND ANOTHER FAMILY ABOUT TO BE SEPARATED.

A SLAVE MARKET IN ATLANTA, GEORGIA, IN THE EARLY 1860S

Family, he tells of the legacy of "the evil institution" upheld by his wealthy white family. The family members owned more than twenty plantations in South Carolina, along the Cooper River north of Charleston.

The family, as he puts it, "was in the slave business, close to four thousand black people born into slavery to the Balls or bought by them." When one of his ancestors, John Ball, was forty-four, he married Caroline, who at nineteen was younger than his two sons by a previous marriage. She soon gave birth to twins, Martha and Caroline, and their father was overjoyed. The right gift to celebrate the event, he thought, would be to give twins to each of his twins. So, calling the family together, he presented each of the infants with a pair of child slaves, each about ten. One set of these slave children was American born; the other set recently imported from Africa.

When a man named John Ball from a different family died at fifty-seven, his will called for the sale of his seven plantations and his 695 slaves. His heirs didn't hesitate to break up families. They heard "the rustle of cash," and couldn't resist it. On one plantation, thirty-nine people were sold to fourteen buyers. On another, thirty-one went to eight buyers, then twenty-three to nine buyers, and so on. Finally, fifty people were placed on the auction block alone, all of them "torn completely from their wives, husbands, children, parents."

NINE

"I Have a System to Destroy"

Blacks never accepted their enslavement. On the ships that carried them in chains to America, they protested in every way possible. Rather than submit, they starved themselves to death, they broke their chains and seized hold of the slave ships, they threw themselves overboard.

After the Revolutionary War, in which many fought, free blacks became active in the antislavery movement. They had heard the words *liberty* and *freedom*, and they wanted those slogans made a reality for themselves and for the blacks still in bondage. They used pulpit, platform, and press in their demand for liberation.

David Walker, freeborn but the son of a slave, became a leader of his people's protest movement in Boston. In 1829, he published his famous *Appeal*, a pamphlet that was an eloquent indictment of slavery. He urged all black people to rise in violent insurrection, if necessary, against their oppressors. Yet he also suggested Christian forgiveness would be offered if slaveholders freed their slaves voluntarily. The *Appeal* quickly reached many Africans who could read, and infuriated the slave masters. In his pamphlet, Walker said:

> Can our condition be any worse? Can it be more mean and abject? If
> there are any changes, will they not be for the better, though they may

Death of Capt. Ferrer, the Captain of the Amistad, July, 1839.

Don Jose Ruiz and Don Pedro Montez, of the Island of Cuba, having purchased fifty-three slaves at Havana, recently imported from Africa, put them on board the Amistad, Capt. Ferrer, in order to transport them to Principe, another port on the Island of Cuba. After being out from Havana about four days, the African captives on board, in order to obtain their freedom, and return to Africa, armed themselves with cane knives, and rose upon the Captain and crew of the vessel. Capt. Ferrer and the cook of the vessel were killed; two of the crew escaped; Ruiz and Montez were made prisoners.

A SCENE FROM THE 1839 REVOLT OF AFRICANS ABOARD THE SLAVE SHIP *AMISTAD*. KIDNAPPED FROM THEIR HOMELAND, THEY WERE TO BE SOLD INTO SLAVERY IN CUBA. THE SHIP WAS SEIZED BY THE U.S. NAVY, BUT IN A HIGHLY PUBLICIZED TRIAL, THE SUPREME COURT RULED THAT ON THE BASIS OF "ETERNAL PRINCIPLES OF JUSTICE" THE MEN MUST BE FREED. THEY WERE RETURNED TO AFRICA.

appear for the worst at first? Can they get us any lower? Where can they get us? . . . .

The Indians of North and of South America—the Greeks—the Irish, subjected under the king of Great Britain—the Jews, that ancient people of the Lord—the inhabitants of the islands of the sea—in fine, all the inhabitants of the earth (except, however, the sons of Africa) are called men, and of course are, and ought to be free. But we (colored people) and our children are brutes! And of course are, and ought to be slaves to the American people and their children forever! To dig their mines and work their farms; and thus go on enriching them from one generation to another with our blood and our tears! . . .

How would they like for us to make slaves of, and hold them in cruel slavery, and murder them as they do us? . . . I ask you, had you not rather be killed than to be a slave to a tyrant, who takes the life of your mother, wife and dear little children? . . . [A]nswer God almighty; and believe this, that it is no more harm for you to kill a man who is trying to kill you, than it is for you to take a drink of water when thirsty. . . . The greatest riches in all America have arisen from our blood and tears. . . . But Americans, I declare to you, while you keep us and our children in bondage, and treat us like brutes to make us support you and your families, we cannot be your friends. You do not look for it, do you? Treat us like men, and we will be your friends.

Some Southern states were so outraged by the *Appeal* that they passed laws forbidding its circulation and made it a crime punishable by death to bring in such literature. A price was offered for Walker's head as his *Appeal* became one of the most widely circulated publications authored by a black.

A few months after the *Appeal's* third printing, Walker was found dead near his clothing store in Boston. It was widely rumored that he had been poisoned.

Antislavery newspapers began appearing as early as 1812. They

OPPOSITION TO SLAVERY BEGAN IN THE COLONIAL ERA AND OVER TIME TOOK VARIED AND COMPLEX FORMS. BY 1839 THE LIBERTY PARTY HAD FORMED IN ORDER TO PRESS THE CAUSE. IT PLAYED AN IMPORTANT ROLE IN THE PRESIDENTIAL ELECTIONS OF THE 1840S. THIS ENGRAVING SHOWS THE LARGE TURNOUT FOR AN ANTISLAVERY MEETING IN NEW HAMPSHIRE IN 1841.

were a strong influence on many whites and blacks, so threatening to the slave power that several presses were wrecked and some of their editors were beaten.

While some abolition papers were shut down overnight, one of them, perhaps the most important, continued its fiery life until slavery was ended. It was the *Liberator*, the voice of William Lloyd Garrison, a Boston printer. "I have a system to destroy," he wrote, "and I have no time to waste."

From its launching on New Year's Day in 1831, the paper became the leading voice of abolition. By 1840, the American Anti-Slavery Society, which Garrison helped found, had 250,000 members, published more than two dozen journals, and had some 15 state organizations with 2,000 local chapters.

Many great men and women of that day, black and white, entered the ranks of reform. They included preachers, teachers, writers, artists, professors, businessmen, farmers, and feminists.

The foremost black abolitionist was Frederick Douglass. Born a slave in Maryland in 1817, as a boy of ten he learned to read from his master's wife until her enraged husband stopped it. Hired out as a shipyard worker in Baltimore with all his pay going to his master, at the age of twenty-one, Douglass managed to escape and reach New York. He became one of the movement's most effective speakers, telling the story of his life in slavery and of his escape to freedom. He was attacked by proslavery mobs several times before he launched his own abolitionist paper, the *North Star*, in 1847 in Rochester, New York. Douglass would never cease speaking, writing, agitating, and organizing for the full freedom of all African Americans. As he had predicted, the Civil War would end slavery, but the racial stereotypes and patterns of racial discrimination it had fostered did not die. He kept fighting those relics of slavery until his death in 1895.

Catching a spark from Garrison's fiery speeches and editorials, the young Lydia Maria Child of Massachusetts became the first American to write a book exposing the roots of racism. *An Appeal in Favor of That Class of Americans Called Africans* called for ending all forms of racial discrimination, from employment and segregated schools to laws banning the marriage of people of different races.

So vivid and powerful was her argument that 130 years later, at the height of the Civil Rights movement, it was reissued as a weapon serving the modern cause. Her *Appeal*, first published in 1833, is an inspiring model of how institutional racism can be cured through an

FREDERICK DOUGLASS'S FAMOUS AUTOBIOGRAPHY WAS PUBLISHED IN MANY EDITIONS IN THE UNITED STATES AND IN EUROPE. HE HELPED MAKE THE STRUGGLE TO END SLAVERY THE CRUCIAL ISSUE OF LIFE IN THE UNITED STATES.

LYDIA MARIA CHILD USED HER PEN BRILLIANTLY TO TORMENT THE COMPLACENT UNTIL THEIR CONSCIENCES WERE AWAKENED TO THE HORRORS OF SLAVERY.

interracial solidarity movement dedicated to making democracy, justice, and equality a reality for all human beings.

When she wrote the *Appeal*, Child was a popular author risking career and comfort for the strenuous and often dangerous life of the crusader. She is one of the many great women of her time never given due recognition for their courageous achievements.

The Trail of Tears

For the Native Americans as well as African Americans, the invention of the cotton gin proved to be a disaster of almost unimaginable proportions. By 1800, southern planters began to move deeper south and west to expand their highly profitable cotton cultivation.

The earlier planters settled on empty land, but before long, they clashed with some of the 60,000 Indians of the region. The largest tribes were the Cherokee and the Creek. They lived in the western Carolinas, Georgia, and eastern Alabama and Tennessee. Other parts of the region—western Alabama, northern Mississippi, and Florida—were the homes of the Choctaw, Chickasaw, and Seminole tribes.

These "Civilized Tribes" (so-called by the whites) all lived in stable farming communities. They had built up valuable plantations, mills, and trading posts, adapting themselves to the whites' way of life. It was in part the effect of the U.S. government's policy of "Christianizing" and "civilizing" the Indian nations. But before long the whites in the region made it plain that they were less interested in Indian "progress" than in removing them entirely from the region and its valuable land.

In the early 1830s, the federal government set out to relocate all the

A PAINTING BY ROBERT LINDNEUX SHOWS THE CHEROKEE, TORN FROM THEIR ANCESTRAL HOMES IN THE SOUTHEAST AND FORCED TO MARCH ON THE LONG TRAIL OF TEARS.

southeastern tribes onto lands west of the Mississippi River. President Andrew Jackson, himself a slave owner, backed the planters' demands that the Indians leave their homelands and migrate west. The Indians protested when four southern states passed laws providing for the use of force to remove them. Appealed to for help, Jackson was blunt. You have no hope of relief from the federal government, he said. If you don't like it, move.

When the Supreme Court ruled that the southern states' actions were contrary to treaties and the Constitution, that victory did little to protect the Indians. Jackson openly refused to honor the Court's decision.

A rising concern for human rights, which also animated the antislavery movement, led to calls for justice for the Indians. Petitions signed by thousands flooded into Congress to protest their removal. To the up-surge of popular support for the rights of the Indians, Georgia's governor George G. Gilmer had a ready answer:

> Treaties were expedients by which ignorant, intractable, and savage people were induced without bloodshed to yield up what civilized people had a right to possess by virtue of that command of the Creator delivered by man upon his formation—be fruitful, multiply, and replenish the earth, and subdue it.

What did that mean? He was saying treaties were merely tricks to get without force what the civilized white man wanted from the savage Indian.

Senator Theodore Frelinghuysen would not accept the arrogance of the white cotton planters for whom the governor spoke. He asked, "Is it one of the prerogatives of the white man that he may disregard the dictates of moral principles, when an Indian shall be concerned?"

It seems that it was.

The Indian Removal Act of 1830 was adopted by Congress and signed by President Jackson. In 1831–1832, 23,000 Choctaw and some Cherokee were pressured into moving. Others who refused to move were transported by force—the Alabama Creek in 1836, the Chickasaw the next year. In 1837–1838, federal troops ripped 15,000 to 20,000 Cherokee from their homes and herded them over the 800-mile (1,300-km) "Trail of Tears" to present-day Oklahoma. A fourth of the men, women, and children died on the way. Most of the Seminole Indians were driven out of Florida between 1832 and 1835. By 1840, only small fragments of the original tribal people remained on their ancestral lands.

As the whites enforced the removal of the southeastern tribes, the borders of the cotton kingdom spread. Businessmen in the North, functioning as brokers in the cotton trade, as well as planters in the South, were prospering from cotton. It is estimated that in 1850 alone, between $100 million and $150 million went into Northern pockets. It gave the profiteers a big stake in cotton and slavery.

ANDREW JACKSON, OWNER OF ABOUT 100 SLAVES WHEN ELECTED PRESIDENT, MADE THE EXPULSION OF THE SOUTHEASTERN INDIANS ONE OF HIS MAJOR GOALS.

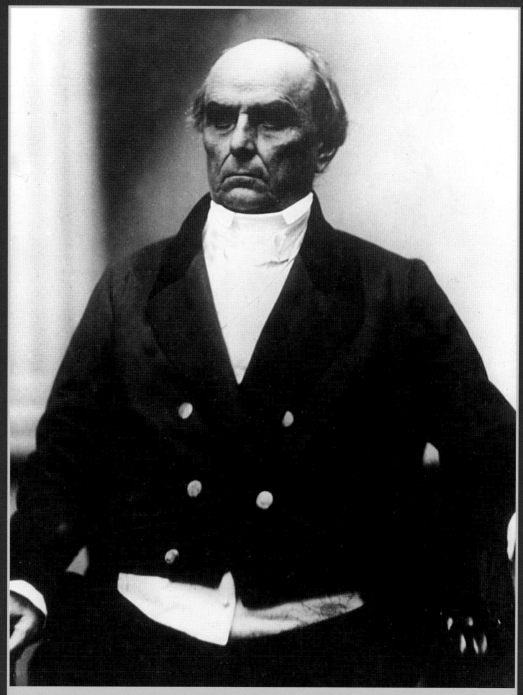

DANIEL WEBSTER, LAWYER AND MASSACHUSETTS SENATOR, WAS GREATLY ADMIRED FOR HIS ORATORICAL POWER. LATER IN LIFE, HE REGRETTED SUPPORTING THE FUGITIVE SLAVE ACT OF 1850, A LAW DESIGNED TO MAKE IT EASIER FOR SOUTHERN SLAVE OWNERS TO RECAPTURE SLAVES WHO HAD ESCAPED TO THE NORTH.

Civil War—and After

Moving into the 1850s, it was clear that America was heading for civil war over the issue of slavery. The slave states were making ever-greater demands upon government to protect their economic and political interests. So righteous did slave masters feel, that they stopped apologizing for slavery as a necessary evil. They began to declare that their "paternalistic" slavery was of positive value to the slaves themselves and morally superior to heartless capitalism. Ever-harsher laws were enacted by the Southern states to keep out abolition literature, to prevent any masters from freeing their slaves, and to further restrict the rights of free blacks.

In their efforts to maintain prosperity, the planters enlisted influential Northerners, such as the noted orator Daniel Webster. A corporation lawyer as well as a U.S. senator, he became their mouthpiece. "The great object of government," Webster said, "is the protection of property at home." By "property" he meant black human beings, the slaves who raised and harvested the cotton.

Now the planters had reason to worry about the supply of slave labor. Remember that in 1808, the slave trade was closed. From then on,

DURING THE CIVIL
WAR, THIS GROUP OF
ESCAPED SLAVES
GATHERED ON THE
SOUTH CAROLINA
PLANTATION OF
CONFEDERATE
GENERAL THOMAS
DRAYTON. WHEN
UNION TROOPS
OCCUPIED THE
PLANTATION, THE
FORMER SLAVES BE-
GAN TO HARVEST
AND GIN COTTON FOR
THEIR OWN PROFIT.

planters had to rely on the illegal import of slaves, mostly sneaked in from Cuba, or on domestic slave breeding. When the overworked fields of the old cotton states began to lose their fertility, the planters often sold off their workforce. Some planters, as we have noted, continued to make money by breeding slaves for the markets of the lower South.

Although cotton and slavery had brought prosperity to the South's upper class, the region lagged far behind the North in industrial and urban development. The South remained overwhelmingly agricultural. And that was what Southern-born whites much preferred. They had prospered for a long time because industrialism in Great Britain and the North was based largely on the mass production of textiles from the raw cotton produced by slave labor. Three factors had made this possible: the invention of the cotton gin, the expulsion of the Indians from huge tracts of land the planters wanted, and the westward movement of free whites and black slaves.

It is worth noting that even though some of the new western lands were not fertile enough for raising cotton, the planters nevertheless hoped to carve new slave states out of them. It would help them maintain their political power in the Congress and Electoral College.

But in the 1850s, the price of prime field hands began to climb, and buyers worried about losing their margin of profit if the cost of slave labor grew too great. On the New Orleans slave market in 1843, a young male slave sold for $600. The price rose steadily until, in 1860, it tripled, reaching $1,800. The effect? The South insisted the African slave trade had to be reopened. It could gain this only by seceding from the Union and making its own laws.

By 1860, the conflict between the Southern planters and the Northern merchants and industrialists seemed irreconcilable. The South's life was dependent upon its greatest crop, cotton, whose value was fixed in the world market. Planters had to reduce the costs of production and handling. They needed more land as well as more and cheaper slaves. The latter could be achieved only by reopening the African slave trade. Southern planters wanted direct free trade with Europe and lower costs for moving and financing cotton.

The North's needs were different. It wanted to protect its industrial production. It wanted the federal government to impose higher taxes on goods coming in from abroad, thus protecting its domestic market. It wanted public credit to build railroads across the northern part of the continent and thus expand its markets. It wanted a steady flow of immigrants for cheap labor in the factories. It wanted to settle the West by opening up the public lands to free homesteaders. The abolitionist crusade had increased its strength by convincing the Northern and Western churches to oppose reopening the African slave trade.

Each side felt strongly that the nation could no longer exist half slave and half free: it had to become all one thing or the other.

In 1860, the young Republican Party adopted a program that met the needs of Northern industrial capitalism. And in the presidential election that fall, the party's candidate, Abraham Lincoln, who campaigned against any extension of slavery, won the White House.

The South felt that it faced disaster. The slave states began to walk out of the Union six weeks after Lincoln's victory. On March 4, 1861, Lincoln took the oath of office. On April 12, the Confederates fired on Fort Sumter in Charleston's harbor. The Civil War had begun.

Slavery was the central issue of the war from the start. It was not advisable, however, for Lincoln to say so. There was too much anti-black feeling in the North, and he had to consider the loyal slaveholders of the border states who had stuck with the Union. He therefore placed enormous stress on Union, on holding the nation together. It was for that cause that he summoned thousands of volunteers to fight.

PRESIDENT ABRAHAM LINCOLN IN A PHOTOGRAPH DATED 1863. ONE OF AMERICA'S GREATEST WRITERS, HIS SPEECHES CONVEY THE HIGHEST IDEALS OF FREEDOM, DEMOCRACY, AND EQUALITY IN SIMPLE YET ELOQUENT LANGUAGE.

The North had great superiority in population and in industrial strength. Northerners had control of the sea and a solid financial base. What was lacking was military skill. Many of the country's most experienced officers joined the Confederacy. Lincoln had only a tiny army that had to be built up out of a mass of raw recruits. It took years of painful experience to learn how to fight, and to win.

This is not the place to examine the details of what turned out to be a much longer and more costly war than anyone anticipated. From the beginning, the abolitionists pointed out to Lincoln and the North that the Union cause would not triumph unless the war was openly fought to end slavery. "It was freedom for all, or chains for all," they said. When Lincoln, thinking the time was ripe, announced the Emancipation Proclamation in 1862, the friends of freedom around the world knew that what the abolitionists had advocated was at last to be realized.

The Union forces were opened to blacks, and they rushed in to fight for freedom. They volunteered even though they suffered unequal pay, allowances, and opportunities throughout the war, having to fight a double battle—against slavery in the South and discrimination and segregation in the North.

When victory was won in April 1865, 180,000 African-American soldiers and sailors had served in Lincoln's army. Another 250,000 had helped the military as laborers. To put an end to slavery, 38,000 blacks gave their lives in battle, a death rate 40 percent greater than the whites'. If their courage needed more proof, there were twenty African-American soldiers and sailors who earned the Congressional Medal of Honor.

As a war measure, the Emancipation Proclamation did its work. But more protection was needed. It was the Thirteenth Amendment to the Constitution (1865) that abolished slavery; the Fourteenth Amendment (1868) that made all persons citizens and guaranteed them "the equal

102

THE 54TH MASSACHUSETTS COLORED REGIMENT—THE FIRST AFRICAN-AMERICAN REGIMENT IN THE U.S. ARMY—CHARGES FORT WAGNER IN SOUTH CAROLINA ON JULY 18, 1863. COMMANDED BY A WHITE OFFICER, COLONEL ROBERT GOULD SHAW, WHO DIED DURING THE ATTACK, THE REGIMENT INCLUDED CHARLES AND LEWIS DOUGLASS, SONS OF FREDERICK DOUGLASS.

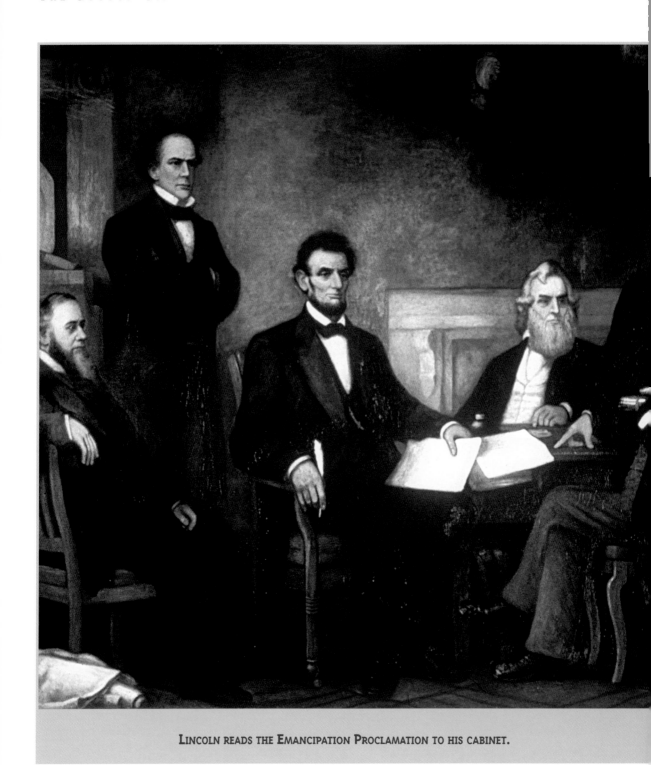

LINCOLN READS THE EMANCIPATION PROCLAMATION TO HIS CABINET.

African Americans in Congress

During the Reconstruction years, fourteen African Americans were elected to the U.S. House of Representatives and two to the U.S. Senate. Most had served in state legislatures or had held state or local office before going to Congress. They not only fought for civil rights for blacks, but were active on behalf of all the leading issues of the time.

Later, black legislators were often blamed for the ills that beset the defeated states in the wake of the Civil War. Biased historians as well as white supremacists leveled false charges against black lawmakers, while ignoring the positive changes they had accomplished in the legislatures in which they had served.

protection of the laws"; and the Fifteenth Amendment (1870) that guaranteed the right to vote for all citizens, regardless of race, color, or previous condition of servitude.

The Union armies had smashed the old order. The slaveholding economy of the South was in ruins.

In what direction would the South move? Would it adopt the politics and principles of the free North? Or would it hold fast to its old way of life?

A CLASSROOM IN
THE POST–CIVIL
WAR SOUTH

Much depended on what the North would do. Northern manufacturers, merchants, bankers, and brokers saw the South as a limitless horizon for expansion and profits. Unlike the abolitionists, their chief interest was in creating calm conditions for investments to prosper, not in guaranteeing full freedom and equality for the emancipated blacks. After an agonizing period of violent resistance in the South to any steps toward that goal, a Reconstruction program was adopted by Congress in 1867, backed by military force. Men and political parties in both the North and South tried to mold that policy. There was fierce controversy between both sides and within both sides over how to do it. Black farmers and workers, preachers, teachers, and politicians who had come out of slavery battled for jobs, education, land, and political power. The era lasted only to 1877—when a deal between Northern and Southern business and political interests ended it.

Reconstruction failed, said Frederick Douglass, because

> Our reconstruction measures were radically defective. They left the former slave completely in the power of the old master, the loyal citizens in the hands of the disloyal rebels against the government. . . .
>
> To the freedmen was given the machinery of liberty, but there was denied them the steam to put it in motion. They were given the uniform of soldiers, but no arms; they were called citizens, but left subjects; they were called free, but left almost slaves.

The central failure, Douglass said, was the refusal of Congress to give the freedmen a chance to obtain good land of their own. Yet those dozen years were not a total loss. The great mass of freed people did see an improvement in their living conditions. They had better homes and furnishings, food, and clothing. They built their own schools and churches and paid for the teachers and ministers. They started their own fire companies, reading clubs, burial and insurance associations, military companies, and fraternal groups.

In that brief span of time, blacks learned how to use freedom. From slaves they made themselves into farmers and businessmen, students and teachers, lawyers and bishops, jurors and judges, sheriffs and senators.

They organized, they learned, they grew, they fought. It was not their fault that they lost. It was a failure of the whole American people. Nor were blacks the only ones who lost. The whole nation suffered a terrible defeat when Reconstruction was abandoned. It set back the cause of freedom and democracy for generations. Not until the Civil Rights movement began in the early 1950s, did hope for full freedom and equality rise again. And though that struggle made gains, the nation has still to free itself from all vestiges of the grim heritage of slavery.

Eli Whitney's invention of the cotton gin played an innocent role in shackling America to the effects of slavery. Neither he nor anyone else could have predicted the enduring influence of the evil institution of slavery upon the lives of every generation since that day.

Further Reading

Collier, Christopher, and James Lincoln Collier. *The Cotton South and the Mexican War, 1835–1850*. New York: Benchmark Books, 1998.

Miles, Lewis. *Cotton*. Vero Beach, FL: Rourke Enterprises, Incorporated, 1994.

Santella, Marc. *Eli Whitney*. Chanhassen, MN: Child's World, Incorporated, 2003.

Web Sites

Cotton Gin

http://sc.essortment.com/cottongin_rciv.htm

The Cotton Gin: Eli Whitney

http://inventors.about.com/library/inventors/blcotton_gin.htm

History of Cotton

http://www.kings.k12.ca.us/central/cuesd.a/tq/ag/history.html

My History Is America's History

http://www.myhistory.org/historytopics/articles/cotton_gin.html

Teaching with Documents: Eli Whitney's Patent for the Cotton Gin

http://www.archives.gov/digital_classroom/lessons/cotton_gin_patent/cotton_gin_pat

Bibliography

Ball, Edward. *Slaves in the Family*. New York: Ballantine, 1999.

Bruno, Leonard C. *Tradition of Technology: Landmarks of Western Technology in the Collection of the Library of Congress*. Washington, DC: Library of Congress, 1995.

Burrows, Edgar G., and Mike Wallace. *Gotham: A History of New York City to 1898*. New York: Oxford University Press, 1999.

Cahn, William. *A Pictorial History of American Labor*. New York: Crown, 1972.

Child, Lydia Maria. *An Appeal in Favor of That Class of Americans Called Africans*. Amherst: University of Massachusetts Press, 1996.

Davidson, Basil. *The African Slave Trade*. Boston: Little Brown, 1980.

Davidson, Marshall B. *Life in America*. Cambridge, MA: Houghton Mifflin, 1951.

Degler, Carl. *The Other South: Southern Dissenters in the Nineteenth Century*. New York: Harper & Row, 1974.

Diamond, Jared. *Guns, Germs, and Steel*. New York: W. W. Norton, 1997.

Gutman, Herbert. *Who Built America?* New York: Pantheon, 1989.

Hacker, Louis M. *The United States: A Graphic History*. New York: Modern Age, 1937.

Huggins, Nathan. *Black Odyssey: The African American's Ordeal in Slavery*. New York: Vintage, 1977.

Jarrett, Derek. *England in the Age of Hogarth*. New Haven, CT: Yale University Press, 1986.

Jones, Jacquelyn. *Work: Four Centuries of Black and White Labor*. New York: W. W. Norton, 1998.

Jordan, Winthrop D. *White Over Black: American Attitudes Toward the Negro, 1550–1812*. Chapel Hill: University of North Carolina Press, 1968.

Landes, David S. *The Wealth and Poverty of Nations*. New York: W. W. Norton, 1998.

Mayer, Henry. *All on Fire: William Lloyd Garrison and the Abolition of Slavery*. New York: St. Martin's, 1998.

Morton, A. L. *A People's History of England*. New York: International, 1968.

National Geographic Society. *Those Inventive Americans*. Washington, D.C.: National Geographic Society, 1971.

Nye, Russell B. *The Cultural Life of the New Nation, 1776–1830*. New York: Harper, 1960.

Stampp, Kenneth M. *The Peculiar Institution: Slavery in the Ante-Bellum South*. New York: Knopf, 1956.

Wilson, Mitchell. *A Pictorial History of American Science and Invention*. New York: Simon & Schuster, 1954.

Yetman, Norman R. *Life Under the "Peculiar Institution": Selections from the Slave Narrative Collection*. New York: Henry Holt & Co., 1970.

Zinn, Howard. *A People's History of the United States*. New York: Harper, 1980.

Index

Page numbers for illustrations are in **boldface**.

abolitionists, 27-28, 82-89, **85,** 99, 101

Adams, John Quincy, **26,** 27-28

African Americans, **29, 42,** 82-84, 86-89, 101-105, **102-103, 106-107,** 108-109

    See also slaves

Amistad, 83

Appeals

    of David Walker, 82-84

    of Lydia Maria Child, 86-89

Arkwright, Sir Richard, **14,** 15, 19

Ball, Edward, 78-81

Ball, John, 81

Brown, Moses, 19

Child, Lydia Maria, 86-89, **88**

children, 16, 19-21, **50, 52,** 54-57, **56, 66**

    poem about, 59

    See also education

civil rights movement, 109

Civil War, 46, 86, 98-101, **102-103**

colonial times, 15

Constitutional amendments, 101-105

copies, pirated, 39, 41-43

cotton

    cultivation of, **24,** 64-67

    history of, **10,** 11-13, **12,** 15, 43-44, 93

    name, 9

    picking, **34-35,** 60, **62-63, 66,** 67-71

    plant description, 9, 13

    types, 13, 15, 33, 35, 39

    uses for, 13, 16

cotton gin, 35-43, **38, 40,** 109

Dale, David, 19

danger, **52,** 55

Douglass, Frederick, 86, **87,** 103, 108

Drayton, Thomas, 97

education, 19, **53,** 54, 57-58, **106-107**

Egyptian cotton, 13

Emancipation Proclamation, 101, **104-105**

energy, 16, 21, 37

England, 15-19, 43-44, 98

factories, 15-16, 19

　　　　See also mills

fire, 41

Freeman, Theophilus, 75-78

Frelinghuysen, Theodore, 92-93

Fugitive Slave Act, 94

Garrison, William Lloyd, 85-86

Greene, Mrs. Nathanael, 33-35

guns, 41

health, 16, 19, 51, 57

housing, 22, 51, **53,** 71

industrialization, 15-21, 41, 98

Jackson, Andrew, 92-93, **93**

Jefferson, Thomas, 28, 41

Lincoln, Abraham, 99-105, **100, 104-105**

Livermore, Thomas L., 54

Lowell, Francis Cabot, 21

Lowell, Massachusetts, 21-23

manufacturing, 41

mass production, 41

Miller, Phineas, 33-35, 39, 41

mills, 16-23, **17, 18, 20**

Native Americans, 90-93, **91**

New Lanark, 19

New Orleans, 44

newspapers, 84-85

northern United States, 16-23, 17, 18, 20, 25-27, 43-44, 99-101, 108

　　　　See also abolitionists

Northup, Solomon, 75-78

Owen, Robert, 19

patent, **40,** 41-43

payment, 22, 48, 51-52, 54, 71

pima cotton, 13

piracy, 39, 41-43

plantations, **24, 62-63,** 63, 65, 68-69

poem, 59

political power, 98

prehistoric times, 11

presidents, 28, 41, 92-93, **93**

　　　　See also Lincoln, Abraham

productivity, 22-23, 41, 43

racism, 86-89, 101
Reconstruction, 105, 108-109
Republican Party, 99

salaries. *See* payment
Sea Island cotton, 13, 15
sewing machine, 48
ships, **83**
Slater, Samuel, **16**, **17**, 19-21
slaves, 25-28, **34-35**, **38**, **42**, 43, 44, **61**,
    62-63, **66**, **72**, **76**, 95-98
    daily life, 63-73
    fugitive, 94, **96-97**
    population data, 60-63
    revolt of, **83**
    See also abolitionists; workers
slave auctions, **74**, 75-81, **79**, **80**
southern United States, **24**, 25-28, **29**,
    43-44, **45**, 60-63, 84, 98-99
    post-Civil war, 105-109, **106-107**
    See also Native Americans; slaves
speedup, 22-23
spinning frame, 15
spinning mule, 16
spooling, 48-51, 55
standardization, 41
Supreme Court, 92

towns, 19, 21-23
Trail of Tears, 90-93, **91**
transportation
    of cotton, **45**
    of slaves, **83**, 95-98

United States. *See* northern United States;
    southern United States

Walker, David, 82-84
Washington, George, 28
waterwheels, 16
Web sites, 113
Webster, Daniel, **94**, 95
westward expansion, 98, 99
Whitney, Eli, **31**, **36**
    as businessman, 39-43
    early life, 30-33
    as inventor, 33-37, 41
women, 22, **42**, 46-48, **47**, **48**, **49**, **53**
workers, 16, **18**, 22-23, **42**, 46-58,
    47, **48**, **49**
    See also children; slaves
working conditions, 22-23, 54-57
World Heritage sites, 19

About the Author

Milton Meltzer has published more than one hundred books for young people and adults in the fields of history, biography, and social issues. He has written or edited for newspapers, magazines, radio, television, and films.

In 2001 the American Library Association honored him with the Laura Ingalls Wilder Award for lifetime contributions to children's literature. Among his other honors are five nominations for the National Book Award as well as the Regina, Christopher, Jane Addams, Carter G. Woodson, Jefferson Cup, Washington Book Guild, Olive Branch, and Golden Kite awards. Many of his books have been chosen for the honor lists of the American Library Association, the National Council of Teachers of English, and the National Council for the Social Studies, as well as for the *New York Times* Best Books of the Year list.

Meltzer and his wife, Hildy, live in New York City. He is a member of the Authors Guild, American PEN, and the Organization of American Historians.

DATE			